Biology: Understanding Respiration

Why I Wrote This Book

Respiration is one of the processes that all living organisms have in common. As such it has to be one of the most important processes in biology; and one that it is always worthwhile to study and understand better.

In this book I have explained the main principles of respiration and included an exam question and answer section with exam tips to help students.

Often people who don't know chemistry well find respiration a difficult topic to grasp; so I have made the effort to explain the chemical terms used and make these easily understandable.

Respiration is sure to come up on a Biology course; so I invite you to read this book to gain understanding of this fundamental topic. I am sure it will help you and I am very happy that it will do so.

Book Contents

A Few Fascinating Facts

- An average human being uses around 500,000 cm³ of oxygen a day and produces around 300,000 cm³ of carbon dioxide.

-Earth is the only planet where as far as we know, oxygen exists in its elemental form. This is because it is produced by plants in photosynthesis.

-An average human produces around 10 Mega-joules of energy per day.

ATP (Adenosine Triphosphate) – the chemical that is made during respiration from ADP (Adenosine Diphosphate) and P (Phosphate). It functions as the cells usable supply of energy for all processes that require it; e.g active transport. ATP is formed according to this equation:
ADP + P + Energy -> ATP
ATP is used according to this opposite:
ATP -> ADP + P + Energy
ATP is then continually broken down and reformed in what is called the ATP/ADP cycle.

Respiration – the process by which a chemical substance (substrate) is broken down in the cell to release energy in the form of ATP.

Aerobic Respiration – respiration that requires oxygen.
The overall equation is:
Glucose + oxygen -> Carbon dioxide + water + energy

Anaerobic Respiration – respiration that does not require oxygen.
The overall equation is:
In animals:
Glucose -> lactic acid + energy
In plants and fungi:
Glucose -> ethanol + carbon dioxide + energy

Respiratory Substrate – this is a chemical that can be broken down by the cell to release energy in respiration. The most common substrates are carbohydrates, fats and proteins; humans can use all three of these substrates.

Mitochondria – a cell organelle that is very important in the process of aerobic respiration. Within the mitochondria the Link

Reaction, the Krebs cycle take place in the matrix and the Election Transport Chain takes place on the folded cristae membrane.

These chemical reactions can only occur when oxygen is present in the cell (i.e. under aerobic conditions). When oxygen is not present the mitochondria will shut down and pyruvate will not enter it.

Substrate Level Phosphorylation – the process by which ATP is generated from ADP through the supply of energy from the exothermic (heat producing) reactions in glycolysis and the Kreb's cycle:

ADP + P + Energy -> ATP

It only produces four ATP's for every glucose molecule, two coming from glycolysis and two coming from the Kreb's cycle (one from each of the two turns of the cycle that occur for each glucose molecules).

Oxidative Phosphorylation – the process which ends with the oxidation of hydrogen by oxygen to form water; and takes place in a number of successive steps, which generate enough energy to phosphorylate ADP to make ATP:

ADP + P + Energy -> ATP

This is a key process as it generates many more ATP from one molecule of glucose than the substrate level phosphorylation that occurs in glycolysis and the Kreb's cycle. Most books give the estimate of 32 ATP's produced from oxidative phosphorylation for one molecule of glucose, out of a total of 36.

Matrix – the fluid part of the mitochondria that is made up of enzymes and reacting compounds involved in the link reaction and the Kreb's cycle.

Respiratory Quotient (RQ) – the equation for RQ is:

RQ = Amount of carbon dioxide produced

Amount of oxygen produced

When a carbohydrate, protein or fat substrate is respired it has a certain RQ. This is because these substrates produce different amounts of carbon dioxide related to the amount of oxygen consumed. The RQ values are as follows:

Carbohydrates ;   RQ = 1
Fats ;   RQ = 0.7
Proteins ;   RQ = 0.8-0.9

When aerobic respiration is taking place the RQ is always 1 or less, but because anaerobic respiration produces carbon dioxide and uses no oxygen, when anaerobic and aerobic respiration are occurring together, the RQ measured may rise some way above 1; this is quite often observed in growing seeds for example.

Respirometer – apparatus used to determine the RQ of a respiring organism through measurement of oxygen absorbed and carbon dioxide produced. The RQ may then be used to work out what the organism is respiring.

Carbon dioxide is absorbed by potassium hydroxide solution and can be measured by the weight gain of the solution. Oxygen usage is measured by the volume decrease in the chamber where the organism is living.

Matrix - the solution of enzymes and substrate molecules in the middle of the mitochondria, where the reactions of the Krebs cycle and the Link reaction take place.

Cristae - the folding of the inner mitochondrial cell membrane. This folding increases the surface area for the reactions of oxidative phosphorylation to take place. The cristae have many proteins, a large number of which are enzymes involved in the electron transport chain.

Glycolysis – this is the series of chemical reactions that breaks down one glucose molecule (with six carbon atoms) into two pyruvate molecules (each with three carbon atoms)
Glycolysis does not require oxygen and does not release carbon dioxide. Overall it produces enough energy to make two ATP molecules per glucose.

The Link Reaction – the chemical reaction that converts pyruvate to acetyl coenzyme A. It has this name because it links the processes of glycolysis and the Kreb's cycle. It occurs in the mitochondrial matrix.

The Oxygen Debt – this refers to the amount of oxygen needed to be inhaled after exercise in order to oxidise the lactic acid that has accumulated due to anaerobic respiration. Athletes may continue breathing at a heavier than normal rate for up to a few minutes after a race.

Coenzyme - an organic (carbon based) molecule that attaches to an enzymes active site and helps it to catalyse a particular reaction by accepting a chemical group (such as a hydrogen atom); it then transfers that chemical group to another, different enzyme and reaction. Examples in the chemical reactions involved in respiration are NAD and FAD which are involved in redox reactions. NAD and FAD are particularly important in the production of ATP under aerobic conditions, as they ferry hydrogen atoms from the Kreb's cycle reactions, to the electron transport chain.
Coenzyme A is involved in the Link Reaction, ferrying the acetate group into the Kreb's cycle.

Reduced NAD/ Reduced FAD – the coenzymes NAD and FAD after they have had two hydrogen atoms added to them – i.e. they have been chemically reduced.

Therefore reduced NAD is actually $NADH_2$ and reduced FAD is actually $FADH_2$.

Redox reaction – this is a chemical term that is very important in understanding respiration. During a redox reaction one chemical has been reduced, while a different chemical has been oxidised. So the term redox means that both REDuction and OXidation have taken place.

Chemically, oxidation happens when a chemical gains oxygen or loses hydrogen atoms or electrons, while reduction is the exact opposite.

Therefore $NADH_2$ is also called reduced NAD because it has been reduced through the gain of two hydrogen atoms.

Respiration overall is a redox process; glucose is being oxidised by gaining oxygen to form water and carbon dioxide, while oxygen is reduced to water during this process.

There are also many individual redox reactions involved in respiration, an example being the steps in the electron transport chain, where electrons are passed between carrier proteins.

The Electron Transport Chain – the series of proteins that are found on the membrane of the cristae in the mitochondria. Hydrogen atoms first of all are passed onto the chain by $NADH_2$ or $FADH_2$. Then hydrogen is passed between a pair of these carriers, generating an ATP.

Then the hydrogen atoms split into an H+ ion and an electron (a hydrogen atom consists of a proton, that is an H+ ion, and an electron). The electrons are passed down further carriers, releasing a further two ATPs. The energy needed to form these ATPs comes from the series of redox reactions that occurs as first, hydrogen, and then the electrons are passed between the carriers, the resulting reduced protein carrier being at each step more stable than the previous one.

The electrons are finally picked up, along with the two H+ ions, by oxygen to form water. Oxidative phosphorylation is the name for the chemical process that occurs along the electron transport chain.

The proteins involved in the chain are all enzymes, which catalyse this series of redox reactions. One of these enzymes is called cytochrome oxidase, and is inhibited by cyanide, hence stopping the chain from working or producing ATP; therefore death occurs rapidly as heart muscle cannot produce enough ATP to beat.

Chemiosmotic Theory - explains how the electron transport chain produces the ATP from ADP and P.

The theory says that as the hydrogen, and then the electron, are passed down the carriesrs embedded in the cristae membrane, the redox reactions generate enough energy to pump H+ ions into the intermembrane space, against the concentration gradient. Therefore H+ ions are concentrated in the intermembrane space and then allowed to flow back across the membrane through the ATPase enzyme (a large stalked particle) which spans the membrane. This flow of H+ ions down the concentration gradient, generates enough energy for the ATPase enzyme to make an ATP from ADP and P.

Evidence for this theory includes a high H+ ion concentration (a low pH) found in the mitochondrial intermembrane space.

Oxidative Phosphorylation – the process by which, first of all a hydrogen atoms (removed from $NADH_2$/$FADH_2$), and then electrons, are passed along a number of carrier proteins. This process generates three ATPs for every hydrogen atom removed from reduced NAD ($NADH_2$) or reduced FAD($FADH_2$). As removal of a hydrogen is termed 'oxidation', and the hydrogen and electron ultimately react with oxygen to form water, this process is oxidative, and produces ATP from ADP and P; hence the term – 'Oxidative Phosphorylation.'

This process (which occurs on the cristae of the mitochondria) releases thirty two ATPs for every glucose during aerobic respiration. This is a much greater amount than the four ATPs released during substrate level phosphorylation.

Oxidative phosphorylation will only occur in the presence of oxygen, as oxygen is needed to accept the electron and H+ ions at the end of the chain to form water.

Substrate Level Phosphorylation – ATP generation from ADP and P that can occur thanks to energy released when the substrate molecule structure is broken down during respiration.

In fact, in aerobic respiration, substrate level phosphorylation overall releases only four ATPs per glucose molecule; a small amount in comparison to the thirty two ATPs produced by oxidative phosphorylation.

Chapter Two - Main Questions and Answers

1. What is the difference between anaerobic and aerobic respiration?

Aerobic respiration involves oxygen and has the general equation:
Glucose + oxygen -> carbon dioxide + water + energy
Anaerobic respiration on the other hand, does not involve oxygen and has the general equation:
(in animals): Glucose -> lactic acid + energy
(in plants): Glucose -> ethanol + carbon dioxide + energy

2. Why does anaerobic respiration occur?

It occurs due to a lack of oxygen supplied to the tissue or organism. This often occurs in animal (including human) muscle, when exercise is taking place, because oxygen demand is higher and the circulation may not be able to supply enough; resulting in anaerobic respiration.
If you have ever had a 'stitch' in your side or a muscle cramp, these are due to the build up of lactic acid formed by anaerobic respiration in the muscle.
Anaerobic respiration in higher organisms is a kind of last resort, which takes place when oxygen supplies are low; but can only be tolerated for short amounts of time; because lactic acid is a toxin. Some lower organisms such as bacteria are only able to respire anaerobically (are 'obligate anaerobes') and are actually killed by the presence or oxygen.

3. How does anaerobic respiration result in the 'oxygen debt'?

Extra oxygen is needed at the end of vigorous exercise to oxidise the lactic acid that has built up in the muscles, due to the anaerobic respiration that occurs during exercise.

Lactic acid needs to be oxidised to carbon dioxide and water, since lactic acid is a metabolic toxin (which is why it causes cramps). It is actually oxidised in the liver.

The oxygen debt can be quite large and for this reason it may take a few minutes for breathing rate to return to normal after exercise.

4. Why does respiration take place?

Respiration, whether aerobic or anaerobic; takes place in cells to provide energy. All the cells in an organism need to respire in order to stay alive. Cells that cannot respire will die; this happens when the circulation is cut off to a limb in an animal, because the circulatory system can no longer supply oxygen to the tissues.

Respiration and ATP occur in all cells. All cells need energy to carry out their functions and therefore they need to respire a substrate to make ATP, which is the universal, usable form of energy.

There would are no living organisms on Earth that do not perform respiration.

5. Where does respiration take place in the cell?

In eukaryotes the answer is that anaerobic respiration takes place in the cytoplasm, while aerobic respiration takes place in both the cytoplasm (glycolysis) and the mitochondria (the Krebs cycle and oxidative phosphorylation).

Eukaryotic cells that have a high energy output, such as those that perform active transport (which requires ATP) in the first convoluted tubule of the kidney, tend to have a large number of mitochondria.

Prokaryotes do not have mitochondria, and respiration takes place on the inside of the cell membrane and in the cytoplasm.

6. *What is the most commonly used respiratory substrate?*

Glucose. This is probably because glucose is the main product of photosynthesis and therefore is in plentiful supply either in as a monosaccharide, or in the disaccharides, maltose and sucrose, or in starch.
In humans carbohydrates form the main energy food in most cultures; the starch found in potatoes, rice and flour is a polysaccharide made of thousands of glucose units joined together. The starch will be digested to glucose before being respired in the cells.
Fats and proteins may also be used as respiratory substrates. They enter the chain of respiratory reactions at various points, e.g. fats and many amino acids (after deamination; removal of the amine group) enter as acetyl coenzyme A, which then joins the Krebs Cycle.

7. *How do we demonstrate that an organism is using a particular respiratory substrate?*

A respirometer is used to determine the amount of carbon dioxide being produced and the amount of oxygen being absorbed by the organism and then the RQ is determined using the equation:

$$RQ = \frac{\text{Amount of carbon dioxide produced}}{\text{Amount of oxygen consumed}}$$

Each respiratory substrate has its own typical RQ when it is being respired:

Carbohydrates ;	RQ = 1.0
Fats ;	RQ = 0.7
Proteins ;	RQ = 0.8-0.9

What often happens though is the result of this kind of experiment is that the RQ comes out as about 0.8 and it is unclear whether the organism is respiring a range of these substrates or whether it is respiring only protein. However, if the RQ is near 1.0 it can be concluded that the organism is mainly respiring carbohydrates, while if it is near 0.7, then fats are the main respiratory substrate.

An RQ of more than 1.0 indicates that anaerobic respiration is taking place, and is often observed with germinating seeds.

8. What happens during aerobic respiration?

The glucose is completely broken down into carbon dioxide and water.

The reactions that take place are usually presented as four different stages, glycolysis, the Link Reaction, the Krebs cycle and the electron transport chain.

Glycolysis takes place in the cytoplasm of the cell, while the other reactions take place in the mitochondria.

9. What happens during glycolysis?

Overall, glucose (a compound with six carbon atoms) is broken down into pyruvate (a compound with three carbon atoms), with the overall production of two molecules of ATP from ADP and P.

No carbon dioxide is produced and no oxygen is consumed. Initially two ATPs are actually used up to turn glucose into fructose 1,6 bisphosphate; this destabilises the glucose molecule so it can be broken down (in chemical terms the two ATPs provide the activation energy needed to let the glycolysis take place).

Eventually though each glucose releases four ATPs in glycolysis, giving an overall yield of two ATP molecules.

Each glucose molecule is broken down into two pyruvate molecules.

During glycolysis two NAD molecules are reduced to $NADH_2$ by gaining hydrogen atoms from the respiratory substrate (this is termed a 'redox' – short for 'reduction/oxidation' reaction). In aerobic conditions these $NADH_2$ (also called 'reduced NAD') molecules will enter the mitochondria and give rise to ATP production). In anaerobic conditions the $NADH_2$ molecules will react with the pyruvate to form lactic acid.

Under both aerobic and anaerobic conditions the end result is the NAD is regenerated from the $NADH_2$ and the regenerated NAD can then receive more hydrogen atoms i.e. react further. Without this regeneration, NAD would run out, and glycolysis would stop.

10. What happens during the link reaction?

The pyruvate formed during glycolysis will enter the mitochondria under aerobic conditions.

In the link reaction, the pyruvate (a compound with three carbon atoms) is converted to acetyl coenzyme A (a compound with two carbon atoms) and carbon dioxide.

This is the first reaction in the series of chemical reactions starting with glycolysis, in which carbon dioxide is released; in fact all the reactions that release carbon dioxide during respiration take place in the mitochondria under aerobic conditions.

It is a redox reaction that forms $NADH_2$ (reduced NAD) from NAD. This $NADH_2$ will then give rise to ATP production in the electron transport chain (see later).

Coenzyme A is involved in this reaction, joining with the acetate product, to form acetyl coenzyme A. When this joins the Krebs cycle, the coenzyme A is released, so that it can take part again in the Link Reaction.

11. What is the significance of the Krebs cycle being a cycle?

First of all, the acetate (a compound with two carbon atoms) from acetyl coenzyme A, joins with oxalacetate (a compound with four carbon atoms) to form citric acid (a compound with six carbon atoms).

It is important to realise that oxaloacetate is ultimately regenerated from the citric acid. That is why the series of reactions forms a cycle, and this is very important as it means that however fast the reactions take place, the cycle will never run out of intermediates, because they are always regenerated. Therefore respiration may take place at very different rates depending on whether the organism is moving rapidly, or growing etc; but the cycle ensures that these important reactions will never have to stop due to a lack of intermediates (which could happen otherwise).

12. What happens during the Krebs cycle?

The acetate is turned into carbon dioxide and hydrogen atoms are removed from it with the help of decarboxylase and dehydrogenase enzymes respectively.

Coenzymes NAD and FAD work along with the dehydrogenase enzymes, picking up the hydrogen atoms to form reduced NAD ($NADH_2$) and reduced FAD ($FADH2$) respectively.

Oxaloacetate is regenerated so the breakdown of acetate is actually all that happens overall.

The carbon dioxide produced is just a waste product, and diffuses out of the mitochondria and out of the cell, and ultimately leaves the organism during breathing (or diffuses out through plant stomata).

The hydrogen atoms are very important because they are ultimately responsible for the production of the majority of ATP that is yielded from one glucose molecule through the process of oxidative phosphorylation, after they are passed down the electron transport chain. The reactions that breakdown the acetate molecule itself in the Krebs cycle only yield one ATP per acetate (per turn of the cycle): as glucose yields two acetates, then only two ATPs result from this substrate level phosphorylation in the Krebs cycle.

The hydrogen atoms removed from the cycle by $NADH_2$ and FADH2 result in the production of thirty ATPs for each glucose.

13. What happens to the Link Reaction, the Krebs cycle and oxidative phosphorylation when oxygen is not present i.e. in anaerobic conditions?

They don't happen.

Oxygen is necessary for oxidative phosphorylation as oxygen acts as the final acceptor for the electrons that are passed down the chain of proteins (and the H+ ions earlier formed from the hydrogen atoms); forming the final stable product: water. If oxygen is not present then it cannot accept the electrons and H+ ions, so the electron transport chain will not work; and so neither will the Krebs cycle, as NADH cannot give up its hydrogen and therefore cannot regenerate NAD. Without NAD the dehydrogenase reactions and enzymes that occur during the Krebs cycle cannot happen, and so the Krebs cycle will not occur.

The Link Reaction will also stop in the absence of oxygen because acetyl coenzyme A will not enter the Krebs cycle, will therefore accumulate, and the Link Reaction will stop.

So, lack of oxygen effectively shuts down the reactions in the mitochondria, resulting in pyruvate accumulating in the cytoplasm. This will then be removed as lactic acid (in animals) and react to form carbon dioxide and alcohol (in plants and fungi) i.e. anaerobic respiration will occur.

14. How does the electron transport chain work?

The electron transport chain consists of a series of protein molecules that are embedded in the cristae (the folded membrane) of the mitochondria. These proteins (which are enzymes) pass, first of all the hydrogen atoms, then electrons between them in a series of redox reactions. At each subsequent protein the electron is in a more stable state, and therefore energy is released; which is ultimately converted to ATP.

The hydrogen is removed from $NADH_2$ and $FADH_2$ and this hydrogen is passed to the first carrier. This is a redox reaction. Then the hydrogen is passed to the second protein in another redox reaction. This reaction generates enough energy to make an ATP from ADP and P.

The hydrogen is then split into an H+ ion and an electron. The electron is then passed on further down the chain of proteins, in another series of redox reactions that results in enough energy being released to form two ATPs from ADP and P. The H+ ions rejoins the electrons at the end of the chain, combining then with oxygen to form water; a very stable product.

Oxygen, then is the final acceptor of the electrons (along with H+ ions) and this final reaction drives the whole electron transport chain as the electron is finally removed in a stable form; allowing the chain of redox reactions to continue passing the electron along.

If oxygen is not present the electron transport chain will not work, and neither will the Krebs cycle and Link Reaction. That is why oxygen is necessary for aerobic respiration to occur; it is a reactive gas that will combine with two electrons and H+ ions to from a very stable product.

The electron transport chain is effectively performing the addition of hydrogen to oxygen to form water (a reaction that releases a lot of energy), in a series of steps, that releases the energy gradually, so that the mitochondria can generate ATP in manageable steps.

For each two hydrogen atoms that pass down the chain, three ATPS are generated. The total amount of ATP generated by the electron transport chain for one glucose molecule is thirty two ATPs (out of a total of thirty six ATPs). We wouldn't live long without the electron transport chain (this is shown by the fatal effect of cyanide poisoning; cyanide inhibits cytochrome oxidase, which is one of the proteins in the electron transport chain). That is why we and other higher organisms, need oxygen constantly, to maintain a high enough rate of ATP production in our cells; without it we would die.

15. What is the difference between the electron transport chain and oxidative phosphorylation?

The electron transport chain is the series of proteins embedded in the cristae, that passes, first the hydrogen atoms, and then the electrons, between them in a series of redox reactions.
Oxidative phosphorylation is the process by which ATPs are generated from ADP and P (thus 'phosphorylation') through the series of oxidative and reductive, i.e 'redox' reactions. This process occurs when the hydrogen atoms, and then the electrons, pass down the electron transport chain.
Oxidative phosphorylation is different from substrate-level phosphorylation, which means the generation of ATP from ADP and P using energy just from the breakdown of the carbon backbone of the substrate molecule.

16. What is the importance of the coenzymes NAD and FAD in the process of respiration?

These coenzymes work along with dehydrogenase enzymes, to help remove hydrogen atoms from the substrate molecules.
In aerobic respiration, they are very important in that they transfer the hydrogen atoms from the matrix of the mitochondria, (where the Krebs cycle occurs), to the cristae membrane. They carry the hydrogen atoms from the matrix to

the cristae, release them there to the electron transport chain, and then return for the next lot of hydrogen atoms again. When NAD and FAD pick up hydrogen atoms from the Krebs cycle the following reaction occurs:

NAD + 2H -> $NADH_2$ (reduced NAD)

FAD + 2H -> FADH2 (reduced FAD)

At the cristae, the opposite reaction occurs, where $NADH_2$ and $FADH_2$ give up the hydrogen atoms they carry, to the electron transport chain. (An $NADH_2$ is generated during glycolysis, which enters the mitochondria under aerobic conditions, and ferries this hydrogen to the electron transport chain).

Under anaerobic conditions, the $NADH_2$ generated during glycolysis will give its hydrogen to pyruvate to form lactic acid in animals, and ethanol plus carbon dioxide in plants and fungi. This reaction regenerates NAD; which is important as this NAD can then go back to the accept the next hydrogen atoms in glycolysis thereby allowing the reactions of glycolysis to continue to produce ATP under anaerobic conditions.

NAD is derived from vitamin B3; the lack of it in the diet leads to the deficiency disease called pellagra. FAD is derived from vitamin B2.

17. *What is the mechanism by which oxidative phosphorylation produces ATP?*

The chemiosmotic theory is the most widely accepted theory as to the mechanism.

As hydrogen atoms and then, subsequently, electrons are passed down the electron transport chain the energy generated from the redox reaction is used to pump H+ ions into the intermembrane space (mitochondria have two membranes and the electron transport chain proteins are in the inner membrane). This process is a form of active transport and as such builds up the concentration of H+ ions in the intermembrane space to greater than that on the inside of the cristae membrane.

The very large ATPase enzyme, (called a 'stalked particle' because of its appearance under the electron microscope), straddles the membrane and when H+ ions are allowed to flow through this enzyme from the intermembrane space, they do so down a concentration gradient and therefore release energy in the process. The ATPase enzyme uses this energy to make ATP from ADP and P.

It follows from this that without ADP and P being present ATP cannot be produced and oxidative phosphorylation cannot occur, and it has been shown that the absence of ADP prevents ATP formation and that subsequent addition of ADP to mitochondria results in ATP formation.

Chapter Three - Exam Questions and Answers

Question 1.
How is the folding of the inner membrane of the mitochondria to form cristae an adaptation to its function? (2 Marks)

Exam Tip: this is a typical structure and function type question. Think of what the folding would increase and how this would help the function that this membrane supports.
Answer: The folding of the mitochondrial membrane increases its surface area (1) and this enables more proteins involved in the electron transport chain, that is found on the cristae, to be present. Therefore, more oxidative phosphorylation to create ATP can take place(2).

Question 2.
Name the metabolic pathway in which glucose is converted to pyruvate, where it occurs in the cell, and state how two ATPs are created when glucose is metabolised by this pathway. (3 Marks)

Exam Tip: this should be fairly straightforward to answer, it is just a matter of recalling the information.
Answer: the metabolic pathway by which glucose is changed to pyruvate is called glycolysis (1). It takes place in the cytoplasm(2). Overall two ATPs are created because while two ATPs are used initially to make fructose1-6 bisphosphate, a further four ATPs are released, giving two ATPs overall (3).

Question 3.

Explain the role of FAD and NAD in both anaerobic and aerobic respiration.(6 Marks)

Exam Tip: this question is asking you to explain: that is the key word. Make sure you mention both anaerobic and aerobic respiration.

Answer: FAD and NAD both function as carriers of hydrogen atoms (1). They are coenzymes that work alongside dehydrogenase enzymes to help remove hydrogen atoms from substrate molecules (2).

In anaerobic respiration NAD, first picks up hydrogen during the conversion of triose phosphate to pyruvate (in glycolysis), forming $NADH_2$ (reduced NAD) and then $NADH_2$ gives this hydrogen to the pyruvate; forming lactic acid (in animals) or ethanol and carbon dioxide (in plants and fungi) (3). In this way NAD is regenerated and glycolysis can continue. The NAD is effectively transferring the hydrogen atoms from triose phosphate to pyruvate to create lactic acid (or ethanol and carbon dioxide). Without this process the reactions would stop and NAD has to be regenerated to allow the continued formation of pyruvate (4).

In aerobic respiration the NAD and FAD pick up hydrogen atoms both from glycolysis, the Link Reaction and the Krebs cycle, also acting as coenzymes with dehydrogenase, to form $NADH_2$ and $FADH_2$ (5). They then transfer these hydrogen atoms to the electron transport chain (which leads to the production of three ATPs) and regenerates NAD and FAD so that glycolysis, the link reaction and the Krebs cycle can continue (6).

Question 4.
What is the role of ATP in living things? Give an example. (2 Marks)

Exam tip: this is a straightforward 'what' question. Write what ATP does and give an example.

Answer: the role of ATP in living things is as an energy currency, it is the immediate form of energy used by the cell (1). One example is in active transport when ATP is used to move a substance against the concentration gradient (2).

Question 5.
How is ATP involved in respiration? (6 Marks)

Exam tip: try to ensure that you make six separate points to cover all the marks.

Answer: first of all, ATP is involved in glycolysis in two ways – firstly, it is used up in the early stages of glycolysis (1), the glucose has two phosphate groups from ATP added to form fructose 1.6 bisphosphate. This is needed to activate the glucose molecule i.e. make it unstable, so it can then be broken down later (2). Secondly, ATP is generated in the later stages of glycolysis as triose phosphate is broken down to pyruvate (3). Overall two ATPs are generated from one molecule of glucose during glycolysis.

In aerobic respiration thirty four more ATPs are created from one molecule of glucose, mostly through the process of oxidative phosphorylation (4). The creation of ATP is the reason for respiration, as a constant supply of ATP is necessary for the organism's cells to function (5).

ATP is created in respiration through the phosphorylation of ADP, using energy produced by respiration (6), according to the equation:

ADP + P + energy → ATP .